DITIGAL MULTIMETER

A simplified guide on the perfect usage of digital Multimeter

I0422841

GEORGE WASHER

Table of Content

CHAPTER ONE

Introduction to Digital Multimeter

A digital multimeter (DMM) is an essential tool for engineers, electricians, technicians, and hobbyists alike, enabling precise measurement and analysis of electrical parameters.

Unlike its analog counterpart, a digital multimeter provides accurate readings displayed on a digital screen, making it easier to interpret measurements. From basic voltage checks to advanced troubleshooting tasks, digital multimeters offer versatility and reliability in various applications across industries.

In this outline, we'll delve into the fundamental components, operating modes, instructions for usage, safety precautions, troubleshooting tips, maintenance practices, and potential advanced features of a digital multimeter.

Understanding how to effectively utilize and maintain this indispensable tool is paramount for ensuring accurate measurements and promoting safety in electrical work.

CHAPTER TWO

Basic Components

Display

The display is the primary
interface of the digital
multimeter, typically an LCD
screen that shows
measurement readings
numerically.

Rotary Dial

The rotary dial, also known as the function selector, allows users to choose the desired measurement mode such as voltage, current, resistance, capacitance, etc.

Input Jacks

Input jacks are ports where the test leads or probes are connected. These jacks are usually color-coded and labeled according to the type of measurement being made (e.g., COM for common/ground, VΩmA for voltage, resistance, and current).

Probes

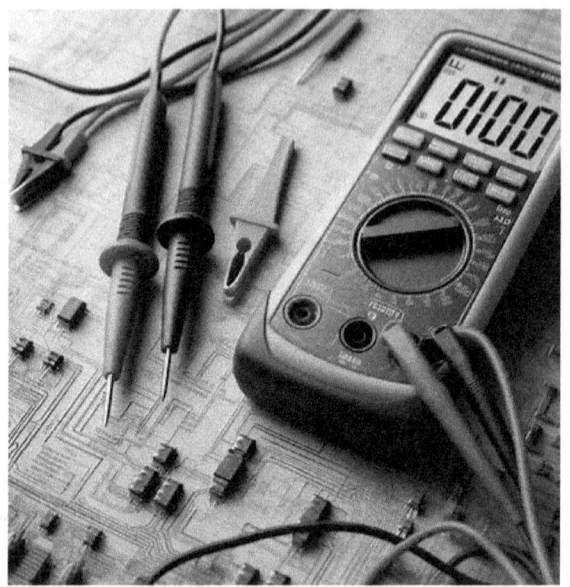

Probes are the detachable accessories used to make contact with the circuit or component being measured. They typically consist of a pointed metal tip and a lead with a connector that plugs into the input jacks of the multimeter.

Function Buttons

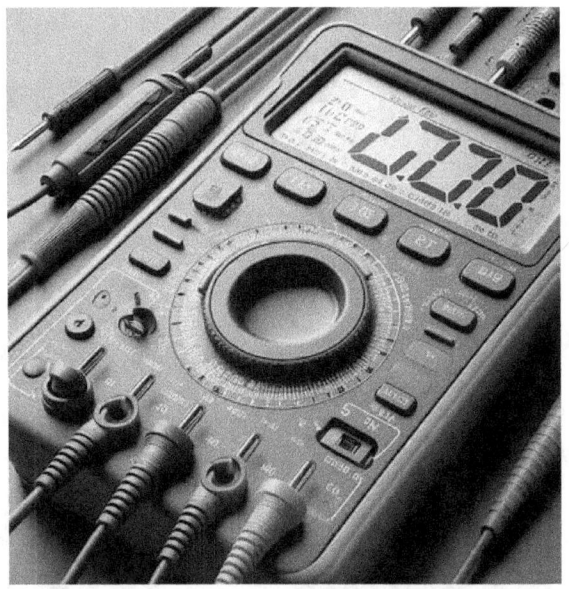

Function buttons allow users to access additional features or settings on the multimeter, such as hold, range, backlight, and mode selection.

Understanding the function and operation of each component is essential for effectively using a digital multimeter in various measurement scenarios.

CHAPTER THREE

Operating Modes

A digital multimeter (DMM) typically offers several operating modes to measure different electrical parameters. These modes are selected using the rotary dial or function buttons. Here are the common operating modes:

1. **Voltage Measurement**

AC Voltage: Measures alternating current voltage.

DC Voltage: Measures direct current voltage.

2. **Current Measurement**

AC Current: Measures alternating current flowing in a circuit.

DC Current: Measures direct current flowing in a circuit.

3**. Resistance Measurement**

Measures the resistance of a component or circuit in ohms (Ω).

4. **Continuity Test**

Checks for the continuity of a circuit by producing an audible beep if there is a low resistance connection between the test leads.

5.**Capacitance Measurement**

Measures the capacitance of capacitors in farads (F) or microfarads (µF).

6. **Diode Test**

Tests the forward voltage drop of diodes and checks for proper functioning.

7. **Frequency Measurement**

Measures the frequency of an AC signal in hertz (Hz).

8. Temperature Measurement

Some advanced digital multimeters come with a temperature measurement mode, usually used with a thermocouple probe.

Each operating mode serves a specific purpose, and understanding when and how to use them is crucial for accurate measurement and troubleshooting in electrical circuits. Users must select the appropriate mode based on the parameter they intend to measure.

CHAPTER FOUR

Operating Instructions

1. Powering On/Off

Turn the digital multimeter on by pressing the power button or rotating the rotary dial to the desired measurement mode.

To conserve battery life, always turn off the multimeter after use.

2. **Selecting Measurement Mode**

Rotate the rotary dial to select the desired measurement mode (voltage, current, resistance, etc.).

Ensure that the selected mode matches the parameter you intend to measure.

3. **Range Selection**

Select the appropriate measurement range based on the expected value of the parameter being measured.

Start with the highest range and gradually decrease to obtain the most accurate reading.

Some multimeters have an auto-ranging feature that automatically selects the best range for the measured quantity.

4. Connecting Probes

Insert the test leads into the corresponding input jacks on the multimeter.

Ensure proper polarity and connection integrity to obtain accurate measurements.

Red probe typically connects to the positive terminal, while black probe connects to the negative or common terminal.

5. Taking Measurements

Connect the test leads or probes to the circuit or component being measured.

Ensure a secure and stable connection to avoid erroneous readings.

Read the measurement displayed on the multimeter's screen.

6. Reading and Interpreting Results

Note the measurement value displayed on the screen.

Pay attention to units (volts, amps, ohms, etc.) and decimal places for accurate interpretation.

Record the measurement if necessary for documentation or analysis purposes.

It's essential to follow these operating instructions carefully to ensure accurate measurements and avoid damage to the digital multimeter or the circuit under test. Familiarizing yourself with the multimeter's functions and practicing proper measurement techniques will improve efficiency and accuracy in electrical work.

CHAPTER FIVE

Safety Precautions

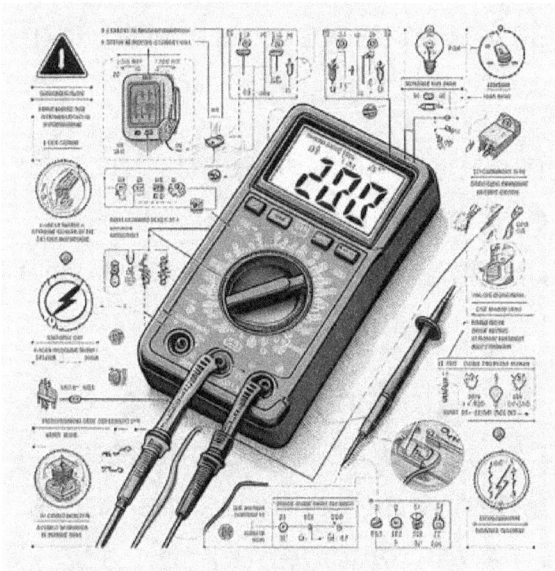

1. Electrical Safety

Always assume that circuits are live and take necessary precautions to avoid electric shock.

Never use the multimeter on high-voltage circuits unless you are qualified to do so.

Use appropriate personal protective equipment (PPE) such as insulated gloves and safety glasses when working with electrical circuits.

Inspect the test leads for any damage or exposed wires before use, and replace them if necessary.

Keep fingers behind the probe's protective barriers when making measurements to avoid accidental contact with live components.

2. Personal Safety

Work in a well-ventilated area to avoid exposure to harmful fumes or gases.

Do not operate the multimeter with wet hands or in damp environments to prevent electric shock.

Keep the work area clean and organized to minimize the risk of tripping hazards or accidental contact with hazardous materials.

Take regular breaks and avoid working while fatigued to maintain focus and prevent accidents.

3. **Device Safety**

Store the multimeter in a dry and secure location when not in use to prevent damage from moisture or physical impact. Avoid exposing the multimeter to extreme temperatures or direct sunlight, as this may affect its performance and accuracy.

Follow manufacturer's instructions for maintenance and calibration to ensure the multimeter remains in proper working condition.

Do not attempt to repair or modify the multimeter unless you are qualified to do so; instead, seek professional assistance if needed.

By adhering to these safety precautions, you can minimize the risk of accidents and ensure a safe working environment when using a digital multimeter in electrical applications. Remember that safety should always be the top priority in any work involving electricity.

CHAPTER SIX

Troubleshooting

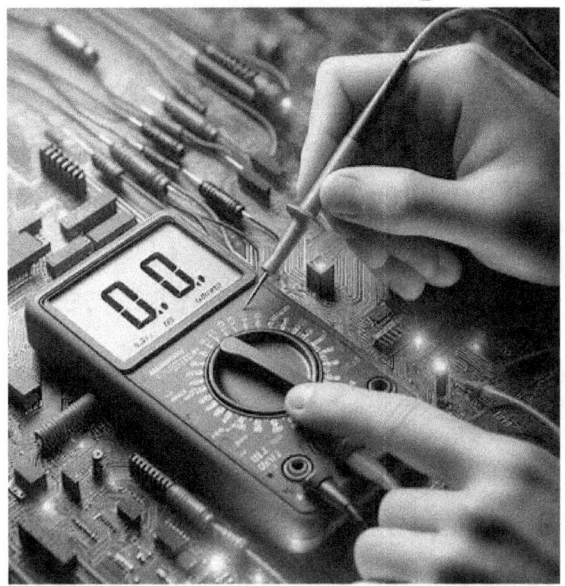

No Power

Check the battery or power source of the multimeter and replace or recharge as necessary.

Ensure proper connections and contacts between the battery and the multimeter.

Check for any blown fuses in the multimeter and replace them if needed.

Inaccurate Readings

Verify that the multimeter is set to the correct measurement mode and range for the parameter being measured.

Check the condition of the test leads for damage or wear and replace them if necessary.

Ensure stable connections between the test leads and the circuit or component under test.

Calibrate the multimeter according to manufacturer's instructions if readings are consistently inaccurate.

Display Issues

Inspect the multimeter's display for any signs of damage or malfunction.

Check the contrast and brightness settings of the display and adjust them if necessary.

Clean the display screen and contacts to remove any dirt or debris that may be affecting visibility.

Intermittent Readings

Check for loose connections between the test leads and the multimeter or the circuit under test.

Inspect the test leads for any signs of damage or fraying that may be causing intermittent contact.

Test the multimeter in different environments to rule out environmental factors affecting the readings.

Fuse Blown

If the multimeter's fuse blows, disconnect it from the circuit immediately to prevent further damage.

Replace the blown fuse with a new one of the same rating and type as specified by the manufacturer

Ensure that the multimeter is set to the appropriate measurement mode and range before reconnecting it to the circuit.

6. Function Button Issues

Clean the function buttons and contacts to remove any dirt or debris that may be affecting their operation.

Check for any physical damage to the function buttons or the multimeter's casing and repair or replace as needed.

If troubleshooting steps do not resolve the issue, refer to the multimeter's user manual for further guidance or contact the manufacturer's customer support for assistance. Avoid attempting any repairs or modifications unless you are qualified to do so to prevent further damage or safety hazards.

CHAPTER SEVEN

Maintenance

Cleaning

Regularly clean the exterior of the multimeter using a soft, dry cloth to remove dust, dirt, and debris.

Use a mild detergent solution and a damp cloth for stubborn stains or spills, but avoid getting water inside the multimeter.

Clean the test leads and probes with a soft cloth or alcohol wipe to remove any dirt or contamination that may affect conductivity.

Inspect the multimeter's input jacks for any debris or oxidation buildup and clean them with a soft brush or compressed air if necessary.

Calibration

Periodically calibrate the multimeter according to the manufacturer's instructions or recommended intervals.

Use a calibrated reference standard or multimeter to verify the accuracy of the measurements.

Adjust the calibration settings as needed to ensure the multimeter provides accurate readings within specified tolerances.

Some multimeters may require professional calibration services, especially for high-precision instruments.

Storage

Store the multimeter in a dry, cool, and well-ventilated environment to prevent damage from moisture, heat, or humidity.

Use a protective case or pouch to store the multimeter when not in use, especially during transportation or storage in a toolbox.

Keep the multimeter away from direct sunlight, extreme temperatures, and corrosive chemicals that may affect its performance or lifespan.

4. **Battery Replacement**

Monitor the battery status of the multimeter and replace the batteries as needed to ensure continuous operation.

Use high-quality batteries recommended by the manufacturer to maintain optimal performance.

Dispose of old or depleted batteries properly according to local regulations and environmental guidelines.

5. **Test Leads and Probes**

Inspect the test leads and probes regularly for signs of wear, damage, or deterioration, such as frayed wires or broken insulation.

Replace damaged or worn-out test leads and probes with new ones to maintain accurate measurements and prevent safety hazards.

Store the test leads and probes properly when not in use to prevent tangling or damage to the cables.

By following these maintenance practices, you can prolong the lifespan of your digital multimeter and ensure accurate and reliable measurements in various electrical applications. Regular cleaning, calibration, proper storage, and maintenance of accessories are essential for maximizing the performance and longevity of the multimeter.

CHAPTER EIGHT

Advanced Features (if applicable)

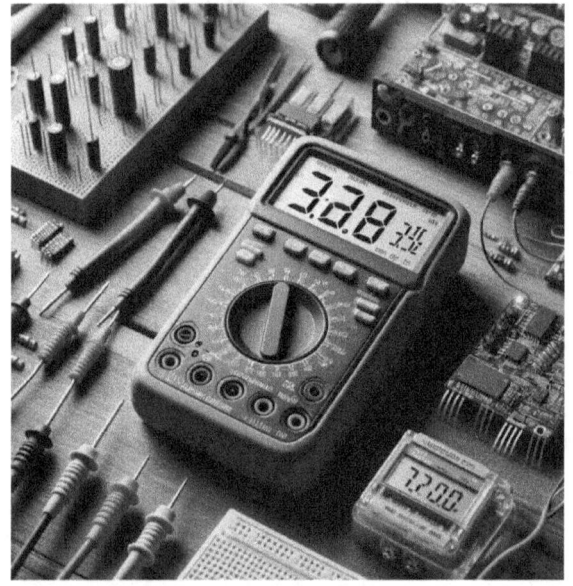

1. Data Logging

Some digital multimeters feature data logging capabilities, allowing users to record and store measurement data over time.

Data logging functionality enables continuous monitoring of electrical parameters and analysis of trends or patterns.

Users can review logged data later for troubleshooting, diagnostics, or performance evaluation.

2. **Connectivity Options**

Advanced digital multimeters may offer connectivity options such as Bluetooth, Wi-Fi, or USB for data transfer and remote monitoring.

Bluetooth or Wi-Fi connectivity allows users to connect the multimeter to a smartphone, tablet, or computer for real-time data sharing and analysis.

USB connectivity enables direct connection to a computer for data transfer, logging, or integration with software applications.

3. **Graphical Display**

Some digital multimeters feature a graphical display with waveform visualization, histograms, or trend graphs for enhanced data analysis.

Graphical displays provide visual representation of measurement data, making it easier to interpret complex waveforms or patterns.

 Users can customize display settings, zoom in/out, and scroll through data for detailed analysis and troubleshooting.

4. **Advanced Measurement Modes**

Advanced digital multimeters may offer additional measurement modes beyond basic voltage, current, and resistance.

Specialized measurement modes may include harmonic analysis, duty cycle measurement, crest factor measurement, and more.

These advanced measurement modes provide insights into the quality and characteristics of electrical signals, particularly in complex waveforms or power systems.

5. **Programmable Features**

Some digital multimeters are programmable, allowing users to customize measurement parameters, settings, and functions.

Programmable multimeters may support scripting languages or programming interfaces for automation, scripting, or integration with test systems.

Users can create custom measurement sequences, automated test routines, or specialized measurement applications tailored to specific requirements.

6. **Environmental Monitoring**

 Certain digital multimeters include sensors or probes for environmental monitoring, such as temperature, humidity, or air quality.

 Environmental monitoring capabilities enable comprehensive testing and analysis of electrical systems in various operating conditions.

Users can assess the impact of environmental factors on electrical performance and reliability, facilitating preventive maintenance and optimization strategies.

These advanced features enhance the versatility, functionality, and performance of digital multimeters, making them invaluable tools for professionals in diverse industries such as electronics, telecommunications, automotive, aerospace, and research. Users can leverage these advanced capabilities for complex measurements, data analysis, automation, and system integration, thereby improving efficiency, productivity, and accuracy in their work.

CHAPTER NINE

Conclusion

Digital multimeters (DMMs) are indispensable tools for engineers, electricians, technicians, and hobbyists, offering precise measurement capabilities and versatile functionality for various electrical applications. From basic voltage checks to advanced troubleshooting tasks, digital multimeters provide accuracy, reliability, and convenience in measuring voltage, current, resistance, and other electrical parameters.

In this guide, we've covered the fundamental components, operating modes, operating instructions, safety precautions, troubleshooting tips, maintenance practices, and potential advanced features of digital multimeters. Understanding how to effectively use and maintain a digital multimeter is essential for ensuring accurate measurements, promoting safety, and optimizing performance in electrical work.

By following proper operating procedures, adhering to safety precautions, and performing regular maintenance, users can prolong the lifespan of their digital multimeters and maximize their utility in diverse electrical applications. Whether performing routine maintenance, diagnosing faults, or conducting complex measurements, digital multimeters serve as indispensable tools for professionals and enthusiasts alike.

In conclusion, digital multimeters represent the pinnacle of measurement technology, empowering users to tackle electrical challenges with confidence, precision, and efficiency.

As technology continues to evolve, digital multimeters will remain essential instruments for navigating the complexities of modern electrical systems and advancing innovation in countless industries.